Development of the Biotechnology Sector: The Iraq Perspective

Sami A. Al-Mudhaffar

Member of Academy of Science (Iraq)
Fellow of Islamic Academy of Science

Introduction

Iraq, which has already entered the biotechnology era, suffer the consequences of the widespread use of such techniques in industrialized countries.

Therefore, the question is no longer whether such a move is desirable, but more how can best be put into use.

Currently people of Iraq are suffering from poor quality of life because of U.N. sanctions, which resulted in:

- **Food shortage and malnutrition**
- **Unsafe drinking water.**
- **Improper sanitation system.**
- **Poor health care.**

Biotechnology alone will not feed Iraqis or give them better health, but it would be irresponsible not to see it as one of the available tools.

This paper:

- **Analyses the major activities of biotechnology in Iraq.**
- **Examines the prospects of strengthening biotechnology in conjunction with conventional technologies.**
- **Finally it outlines the strategies for the promotion of biotechnology in Iraq.**

Historical Aspects of Biotechnology in Iraq

Biotechnological activities were known to ancient Iraqis in Sumer and Babylon.

For instance:

- Fermentation processes and perfume extraction were very well-known to them.
- The first experiment in scientific research was performed in the palace of the king of Babylon.
- The experiment was carried out by two palace ladies who prepared perfumes by distillation from plants (steam distillation).
- Cheese making is believed to have started in the Tigris-Euphrates valley (in present day Iraq`), a approximately 8,000 years ago.

In the early 1970s˙ scientific infrastructure construction began and several research activities were planned and setup.

- Most of these biotechnologal activities were limited to traditional methods to serve their needs. i.e. industrial fermentation, soil microbiology and bioconversion of waste products .
- The government showed its interest in providing support to biotechnology by offering to host the first Arab Conference on Genetic Engineering in 1984 .
- The council of Scientific Research established "Genetic Engineering Center", which became responsible for all research and development in this field.
- The agreement that Iraq established an affiliation of the International Center of Genetic Engineering (ICGEB) with the Scientific Research Council, as the principle Liaison Institute.

During early 1990s:

- Research in agricultural biotechnology dealing with production of Biocides and Biofertilizer and Tissue Culture Technique were carried out by (IPA) Center.
- The Ministry of Higher Education in the middle of 1990s', established a new centers for Genetic Engineering, involved in research on biotechnology.
- The main research objective on biotechnology of these centers involved the application of basic genetic engineering particularly in the following fields:
  - Molecular biology
  - Cell biology
  - Microbial genetics

- Iraq through a new council of biotechnology at the University of Saddam is:
  - Monitoring international development in biotechnology some of these techniques
  - Developing plans for biotechnology and genetic engineering in the sectors of health, agriculture and basic biotechnology
  - Strengthening of research and technology transfer facilities
  - Establishment of international and regional collaboration in the field of biotechnology

Approaches of Biotechnology in Iraq

Most biotechnological activities, in Iraq, are:
- **Fermentation.**
- **Antibiotic industry.**
- **Single cell protein.**
- **Plant biotechnology**

  - Tissue culture
  - Soil fertility, through biological activities Single cell protein.
  - Increasing food production through plant cell-culture
  - Bioconversion of waste for food and feed ingredients

Fermentation
- traditional fermentation industries,

  - Baker's yeast production
  - Ethanol
  - Acetic acid
  - Acetone
  - Butanol
  - Citric acid production

- **In 1970, a factory for making bakers yeast from sugar-beet molasses, was established with plans for the production of compressed yeast.**
- **There are abundant sources of raw materials for fermentation in Iraq. Large quantities of hydrocarbons, and carbohydrate by-products (molasses) and lignocellulose waste are found.**
- **Fermentation of food crops like dates does not receive sufficient attention.**

Antibiotics Industry

- The antibiotics industry in Iraq started in 1970 for the production of penicillin and tetracyclines.
- This industry was discontinued for economical and technical reasons as a result of sanctions.
- The involvement of research and development programs have positive effect for restarting bioindustry and production of different types of antibiotics.
- The pharmaceutical industry in Iraq is directed to satisfy the local market on satisfying the market needs.

Single-Cell Protein

- A research and development program for single-cell protein (SCP) production at pilot plant level started in Iraq in 1982 This industry was discontinued for economical and technical reasons as a result of sanctions.
- The research plan included an economic feasibility study and the assessment of technological and nutritional aspects of SCP under local conditions.
- Another pilot plant was established to utilize date syrup for the production of bakers yeast and SCP.
- The most important achievements of this program were the establishment of pilot research facilities, training of personnel, the nutritional assessment of available commercial SCP products

Plant Biotechnology

- Plant biotechnology applications in Iraq include soil fertility using yeast strains in a mixed culture and cell conversion.
- Scientists have carried out research on nitrogen fixation by grain legumes.

Tissue Culture

- Tissue culture laboratory was established in 1979 at the Agriculture and Water Resources Research Center in Baghdad.
- Another laboratory was initiated at the Genetic Research and Biotechnology Scientific Research Center for the improvement of plant production.
- Tissue culture laboratories were also established at the Universities of Basrah and Mousl and also at the Ministry of Agriculture.
- Date palm propagation by tissue culture was implemented at research institutes and universities in Baghdad and Basrah.
- Other species such as lettuce and potatoes were propagated by tissue culture at the department of Biology of Mosul University.

Other activities

- Research at the Faculty of Agriculture and Biology in the Nuclear Research Center was concentrated on the study of a sexual embryoensis techniques and the induction of mutations by mutagenic agents.

The Situation of Biotechnology in Iraq, Since the Imposition of Sanctions

As a result of sanctions:

- Many fields in public health system are affected sector by sector including critical areas such as:
  - Food security
  - Nutrition
  - Water resources
  - Women's health
  - Children health
  - National health emergencies
  - Hhospital care
  - Humanitarians donations
  - International cooperation.
  - Scientific fields4 (Oncology, Cardiology, Nephrology, Endocrinology, Ophthalmology).
  - Diagnostic Testing and Protection of Blood Supply and Scientific Information and medical education, pharmaceutical and biotechnology inputs
- In agriculture, the U.N. sanctions:
  - Ban the importing of fertilizers
  - Shortages in production of crops

  Led to:
  - The deterioration in the Iraq's populations
  - Nutritional intake (daily caloric intake).

Food Security

- Iraq is facing the following three main threats: food supply, health improvements and environmental protection.
- Food security in Iraq is unique; it is not related directly to biotechnology problems. Genetically altered seeds are not necessarily needed to feed Iraq
- This view rests on two critical assumptions, which we question:
  - The First: is that poverty is not due to a gap between food production and growth of population.
  - The Second: is that biotechnology is not the only or best way to increase agriculture production 0

Applications of Biotechnology in Iraq in the Experimental Stage

Animal production

- The use of biotechnology in animal production in Iraq has occurred, in the field of:
  - Reproduction
  - Animal health
  - Feeding
  - Nutrition
  - Growth and production

- In the field of reproduction, new biotechnologies such as:
  - Embryo transfers
  - In vitro fertilization
  - Cloning
  - Shortages in production of crops

- Animal health can be improved with new biotechnology methods at experimental stage of:
  - Diagnosis
  - Prevention of animal diseases
  - Control of animal diseases

- Biotechnology in animal nutrition concentrates on:

  - Improvement of feed
  - Enzymatic treatment
  - Decreasing of the antinutritional factors

Plant production

- At present, more traditional aspects of biotechnology such as the followings are used:

  - Tissue culture
    This technology was applied in Iraq to improve local varieties of food,

  - Pest and weeds
    Since 1970s' several herbicides were used to control the weeds of corn, cotton and vegetable fields.

Biocontrol Programs Used for Controlling Pests

- Researchers started to evaluate herbicides since 1965, and mid of seventies, herbicides were applied to control the weeds in wheat, rice, corn, cotton, potatoes and tomatoes.
- Increased support is needed to expand research designed to develop new herbicides
- Several centers are involved in biocontrol programs such as:

  - Agricultural and Biological Research Center (Baghdad).
  - State Board for Agricultural Research / Ministry of Agriculture Baghdad.
  - College of Agriculture / Baghdad and Mosul Universities.

- Commercially, research centers introduced two biocontrol mutant fungi. were applied to:
  - Control plant parasitic nematodes
  - Soil born fungi on vegetables and citrus.

- State Board for agricultural research successfully adopted several control measures on:
  - Insect growth regulators
  - Biocontrol agents
  - Fungi
  - Plant extracts

- At the present time, the U.N. sanctions which have been imposed on Iraq since 1990 .
  - Destroyed most of biological control programs
  - Iraq is facing lack of well trained personnel and shortage in facilities.
  - The first generation biotechnologies used in Iraq such as insect resistance, herbicides resistance are not easy to address any more.

Health Biotechnology

- Several centers are interested in carrying on research on health biotechnology. The following may be mentioned.
  - Saddam Center for Cancer and Medical genetics Research (SCCMGR).
  - Institute of Biotechnology and Genetic Engineering for Graduate Students
  - University of Baghdad
  - Department of Genetic Engineering, College of Science, University of Baghdad.
  - Genetic Engineering Departments in a number of Universities.

- SCCMGR is engaged in several lines of biotechnological activities in health:
  - Cloning of tetanus toxic gene into tumor cells.
  - Preparation of tumor cell lines in vitro for gene therapy technique.
  - Preparation of restriction enzymes vectors for gene therapy.
  - Studies on disorders of mitocondrial DNA in muscular dystrophy.

- Several biochemists have participated in various research projects that deal with diagnosis, and monitoring of several types of tumors

Future Biotechnology

Development of biotechnology applied to food and health to face basic human needs in Iraq

- The need for the application of biotechnology to face the basic needs regarding food and health in Iraq is real.
- There are different approaches such as the development of plant biotechnology, biotechnology applied to livestock production and biotechnology applied to food processing.
- The suggested components of biotechnology plan include:

  - Reproduction
  - Micropropagation: through e.g. tissue culture for multiplication
  - Genomics: the molecular characterization of all species
  - Bioinformatics: the assembly of data from genomics analysis into accessible forms
  - Diagnostics: the molecular characterization and identification of pathogens
  - Molecular breeding: the identification and evaluation of desirable traits in breeding programs with the use of marker assisted selection
  - Transformation: the introduction of single genes conferring potentially useful traits into crops, livestock, fish and tree species
  - Vaccine technology: use of modern immunology to develop recombination DNA vaccine for control of lethal diseases.

- Plans for future biotechnology research should be formulated through several priorities:

  - Food security
  - Increase and improvement of agricultural production
  - Production of pharmaceuticals for the extraction of biologically active plant substances
  - Immunology: Production of vaccines and monoclonal antibodies.
  - Use and recycling of agricultural products for the production of ethanol, acetone, butanol and methanol.

Food security

- The applications in agricultural biotechnology in Iraq have the promise of bringing about the much-needed requirements in agricultural production, such as:
  - Carrying resistance / tolerance to abiotic stresses (drought and salinity)
  - Providing options for better rotation to conserve natural resources.

- Use of functional genomics to address complex traits, marker assisted breeding to ensure presence of key genes, improving nutritional quality and managing natural resources better by use of efficient monitoring tools.
- Iraq must be an active participant in this area so that specific needs of food security are achieved.

Agricultural production

- The following classes of agricultural biotechnology are suggested to be used in Iraq in the future:

  - Gene transfer technologies, which provide transgenic plant, resistance to many pests pathogens, herbicide as well as resistance to stress such as temperature, drought and salinities.
  - Non transgenic biotechnological approaches for improving the efficiencies and effectiveness of conventional plant breeding methods.
  - Technologies for better monitoring of natural resources and environment.

- **Additional suggestions for the implementation of the plant are the following:**

Plant Biotechnology

- The establishment of biological treatment plants for sanitary wastewater.
- The establishment of the commercial production of inoculants such as Rhizobium
- Increasing protein content of rice by the application of biotechnological techniques.
- The creation date-palm clones resistant to disease, and the application of tissue culture to improve date-palm varieties.
- Production of secondary metabolites, by tissue culture.

Animal Biotechnology

Biotechnological application in livestock and fish production and the adoption of embryo culture to improve local animal breeds through embryo transfer technology, pre-implantation and embryo freezing

Microbial Biotechnology

- Microbial biotechnology for ethanol production from sugar by-products and methanol production from Agro-industrial wastes.
- Microbial genetics: elimination or degradation of pollutants transformation of cellulolytic nitrogen fixers.
- Proper technology to convert biomass into biofue.
- Bioconversion of lignocellulosic wastes to protein-enriched fermented materials.
- The use of bacterial treatment for the removal of oil and chromium

Health Biotechnology

Medical Biotechnology in Iraq, as a whole, will demand a strong collaboration in order to reduce present gap between developed countries and Iraq to achieve this aim

- A center of bone marrow transplant should be established in one of the hospitals.
- Research projects in gene therapy should be initiated

Pharmaceutical industry

The pharmaceutical industry in Iraq should be expanded and developed so as to meet at least the local requirements. Biotechnological techniques should be introduced.

Environment

- ❑ Applications of natural occurring organisms (e.g. yeast, fungi and plants) should be used to convert hazardous substances in soil.
- ❑ Using microorganisms pollutants from sewage systems to clean up industrial sites.

Forensics

Applications of technologies to forensic science.

Bioinformatics

Should be used for measuring and monitoring thousands of genes at one time.

Cooperation with International Agencies

- Cooperation with Islamic and International agencies and countries are required.
- Well trained scientists from Arab and Islamic countries are also required.
- Post-graduate short training courses sponsored by international organizations such as the United Nations Educational, Scientific and Cultural Organization (UNESCO) should be organized.
- Training of medical personnel in bone marrow transplantation.
- Strengthening capabilities/developing projects/visits/ and training programs of mutual interest to all participating countries in the following areas of biological control:
  - Exchange of biotic agents on a case to case basis.
  - Mass production of host insects and natural enemies.
  - Biological suppression of crop pests by developing joint projects.
  - Computerization of information and networking research organization in different countries.
  - Training in different aspects of biological control.

Establishment of Islamic Biotechnology Center

On the light of what has been presented in this paper I believe that, one would come to the conclusion, that an Islamic Biotechnological Center is really needed. Islamic countries which have longer experience in biotechnology could greatly help in establishing such a center.

Ethical issues related to biotechnology

The Islamic world needs to have sharp opinions on various current issues related to biotechnology and genetic engineering such as genomics, human cloning and genetically modified organisms.

Evaluation of Expression of Proteinic Factors, P53, P21 and Scatter Factor In Benign and Malignant Breast Tumors

Sami A. Al-Mudhaffar

The host information of breast tumor patients and healthy control studies.

Groups	Number	Age	Type of tumor
Malignant breast tumor	28 22	38.3 ± 4.8 41.5 ± 5.7	Infiltrating ductal carcinoma ductal carcinoma in situ
Benign breast tumor	20	34.7 ± 7.5	15 Fibrocystic changes 8 fibrosis 7 adenosis 5 fibro epethilial tumors (fibro adenoma)

Detection of estrogen receptor using Immunohistchemical technique

Portion of tissue specimens that had been fresh frozen in liquid nitrogen were prepared as serial (4-6 mm) slides, then immersed in "Retrival solution" citra, AR10 in microwave for 20 minutes and then were cooled at room temperature for 30 min. After that they rinsed in phosphate-buffered saline solution pH 7.6 and placed in path for 2-5 minutes, the excess of buffer were tapped and wiped around sections by gauze pad. Then the slides were treated with (2%) normal goat serum in phosphate buffered solution for 15 min in a humidified chamber to reduce the non specific binding antibody. After incubation with primary antibody ER88 in phosphate-buffered saline solution for 30 min at room temperature, then washed in phosphate-buffered saline solution for 5 min, then incubated with bridging antibody (goat anti rat immunoglobulin in phosphate-buffered saline solution for 10-20 min.) the slides were washed in phosphate-buffered saline and placed in bath for 2 min, peroxidase-antiperoxidase complex (horse radish peroxidase and anti horse radish peroxidase in phosphate-buffered saline solution) was applied to the sections for 10-20 min, after washing in phosphate-buffered saline for 2 min, the slides were then flooded with diaminobenidin/hydrogen peroxidase in phosphate-buffered saline for 5-15 min.

The slides were washed with distilled water, and then were stained with counter stain haematoxylin 1% for (1-2 min). Finally, the sections were rinsed in gently running tap water for 5 min and dehydrated in serial alcohol 70% → 80% → 95 → absolute ethanol → xylene then cover glasses were applied.

Control slides of estrogen receptor-rich (FG-368M breast carcinoma cells were incubated with primary or control antibody).

Detection of Scatter Factor using Immunohistochmical staining

Portion of tissue specimens that had been fresh frozen in liquid nitrogen were prepared as serial (4-6 μm) slides, then they immersed in "Retrival Solution" citra AR10 in microwave for 20 minutes and then were cooled at room temperature for 20 min. After rinsing in phosphate-buffered saline solution pH 7.6 and placing in bath for 2-5 min, the excess of buffer was tapped and wiped around sections by gauze pad. The slides were treated with (2%) normal goat serum in phosphate-buffered saline solution for 15 min. in a humidified chamber to reduce the non specific binding of bridging antibody. After incubation with primary antibody polyclonal antiserum against recombinant human SF (sheep α-rh SF) in phosphate buffered saline solution for overnight at 4°C. The slides were washed in phosphate-buffered saline solution pH 7.6 for 5 min. After incubation with bridging antibody (goat anti- rabbit immunoglobulin in phosphate- buffered saline solution pH 7.6 for (10-20 min), the slides were washed in phosphate- buffered saline and placed in bath for (2 min). Peroxidase-anti peroxidase (horse radish peroxidase and anti horse radish peroxidase in phosphate-buffered saline solution) were applied to the sections for 10-20 min, after washing in phosphate-buffered saline solution for (2 min), the slides were then flooded with diaminobenziidin / hydrogen peroxidase in phosphate-buffered saline solution for 5-15 min. The slides were washed with distilled water, then were stained with counter stain haematoxylin 1% for 1-2 min. Finally the sections were rinsed in gently running tap water for 5 min and dehydrated in serial alcohol 70% → 80% → 95% → Absolute ethanol → xylene the cover glasses were applied.

Control non immune sheep sera were performed to verify the specificity of staining for scatter factor.

Detection of P$_{53}$ Protein using Immunohistochemical technique

Portion of tissue specimens that had been fresh frozen in liquid nitrogen were prepared as serial (4-6 μm) slides, then they immersed in "Retriveal solution" citra, AR10 in microwave for 20 minutes. and then were cooled at temperature for 30 min. After that they rinsed in Tris-based buffer pHIO solution and placed in bath for 2-5 minutes, the excess of buffer was tapped and wiped around sections by gauze pad. The slides were treated with (2%) normal goat serum in Tris-based buffer pH 10 for 15 minutes in a humidified chamber to reduce the non specific binding of bridging antibody. After incubation with monoclonal antimouse immunoglobulin IgG raised against P$_{53}$ in Tris-based buffer pH 10 solution for 2 hours at room temperature. The slides were washed in Tris-based buffer pH 10 for 5 min. After incubation with bridging antibody (goat antirabbit immunoglobulin in Tris- based buffer solution pH 10 for (10-20 min), the slides were washed in Tris-based buffer pH 10 solution and placed in bath for 2 min. peroxidase- antiperoxidase (horse radish peroxidase and antihorse radish peroxidase in Tris-based buffer pH 10 solution was applied to the sections for 10-20 min, after washing in Tris-based buffer pH 10 solution for 2 min, the slides were then flooded with diaminobenzidin / hydrogen peroxidase in Tris-based buffer pH 10 solution for 5-15 min. The slides were washed with distilled water, then were stained with counter stain for nuclei "methyl green stain" for I-2 min. Finally the sections were rinsed in gently running tap water for smin and dehydrated in serial alcohol 70% 80% 95% xylene, the cover glasses were applied.

Control non immune mouse sera were performed to verify the specificity of staining for P$_{53}$.

Detection of `P$_{21}$ protein using Immunobistochemical technique

Portion tissue specimens that had been fresh frozen in liquid nitrogen were prepared as serial (4-6 μm) slides, then they immersed in "Retrival solution" citra, AR10 in microwave for 20 minutes and then were cooled at room temperature for 30 min. After that they rinsed in Tris-based buffer pH 10 solution and placed in bath for 2-5 minutes, the excess of buffer was tapped and wiped around sections by gauze pad. The slides were treated with (2%) normal goat serum in Tris-based buffer pH 10 for 15 minutes in a humidified chamber to reduce the non specific binding of bridging antibody. After incubation with monoclonal antimouse immunoglobulin IgG raised against P$_{21}$ in Tris-based buffer pH 10 solution for 30 min at room temperature. The slides were washed in Tris-based buffer pH 10 for 5 min. After incubation with bridging antibody (goat antirabbit immunoglobulin in Tris-based buffer pH 10 solution for (10- 20 min), the were washed in Tris-based buffer pH 10 solution and placed in bath for 2 min. Peroxidase- antiperoxidase (horse radish peroxidase and antihirse radish peroxidase in Tris-based buffer pH 10 solution was applied to the sections for 10-20 min, after washing in Tris-based buffer pH 10 solution for 2 min, the slides were then flooded with diaminobenzidin / hydrogen peroxidase in Tris-based buffer pH 10 solution for 5-15 min. The slides were washed with distilled water, then were stained with counter stain for nuclei "methyl green stain" for 1-2 min. Finally the sections were rinsed in gently running tap water for 5 min and dehydrated in serial alcohol 70% 80% 95% Absolute ethanol xylene, the cover glasses were applied.

Control non immune mouse sera were performed to verify the specificity of staining for P$_{21}$.

Nuclear immunohistochemical staining for ER in infiltrating diictal carcinoma tissue section.

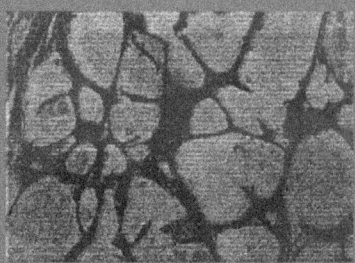

Immunohistochemical staining/or SF protein in ductal carcinoma in situ tissue section.

Estrogen receptors of patients with breast lesions and normal Controls.

Type of lesions	ER+	ER-
Infiltrating ductal carcinoma	24	4
Ductal carcinoma in situ	18	4
Benign	7	13
Normal control	1	24

Expression of SF of patients with breast lesions and normal controls.

Type of lesions	SF(+)	SF(-)
Infiltrating ductal	20	8
Ductal carcinoma in situ	15	7
Benign	4	16
Normal	4	21

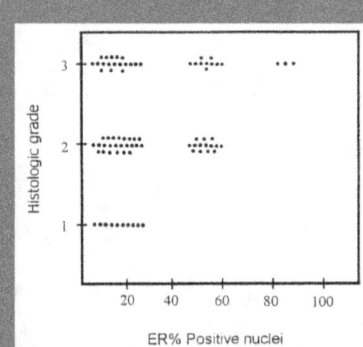

Relationship between the histohgical grade andEstrogen receptor in Infiltrating ductal carcinoma and ductal carcinoma in situ.

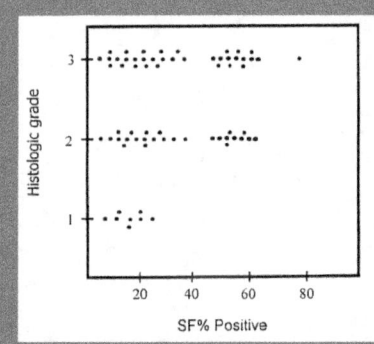

Relationship between the histological grade and Scatter factor expression in Ductal carcinoma in situ.

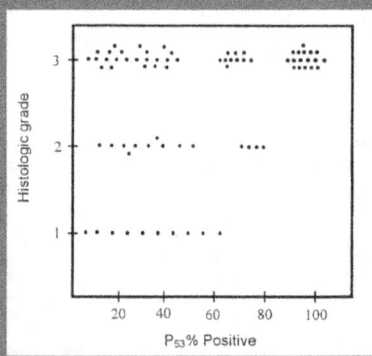

Relationship between the histological grade and the P_{53} Protein expression in Infiltrating ductal carcinoma and ductal carcinoma in situ "Details are explained in the text".

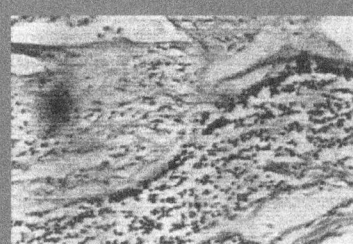

Nuclear immuonohistochemical staining for P_{53} in a poorly differentiated ductal carcinoma in situ tissue section.

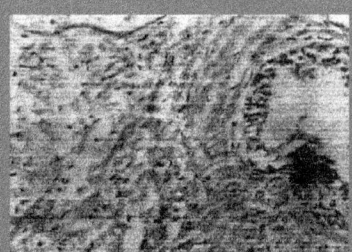

Nuclear immunohistochemical staining for P_{21} in infiltrating ductal carcinoma tissue section.

Expression of P_{53} protein of patients with breast lesions and normal Controls "Details are explained in the text"

Type of lesions	P53 protein expression		
	Neg (-)	Weak (+)	Strong (+)
Infiltrating ductal carcinoma	2	9	17
Ductal carcinoma in situ	3	5	14
Benign	20	0	0
Normal	25	0	0

Expression of P_{21} protein of patients with breast lesionsand normal Controls "Details are explained in the text"

Type of lesions	Expression of P_{21} protein *		
	Neg (-)	Weak(+)	Strong (+)
Infiltrating ductal carcinoma	13	14	1
Ductal carcinoma in situ	7	15	0
Benign	18	2	0
Normal	25	0	0

www.ingramcontent.com/pod-product-compliance
Lightning Source LLC
Chambersburg PA
CBHW080631180526
45168CB00007B/3125